The Lotus Collection

Antique Asian Textiles

A Book of Days

Amber Lotus

Amber Lotus

A Division of Dharma Enterprises

1241 21st Street
Oakland, CA 94607
(415) 839-3931

INTRODUCTION TO THE LOTUS COLLECTION BOOK OF DAYS

The textiles illustrated in this calendar represent some of the finest examples of design and craftsmanship ever created by silk weaving. By virtue of beautiful colors, use of exotic materials, such as gold threads and peacock feather filaments, high quality drawing, and advanced weaving techniques, these textiles are individual works of art. They also stand for historic documents of Chinese and Japanese ancient cultures. Made either for clothing or furnishings, most of these pieces were created for use by either the nobility or top ranking civil officials of each country.

These textiles also provide an intimate view into the lives of these Asian cultures. What we wear and what we choose to put in our homes are very personal statements about ourselves and about the society we live in. Another dimension is added to these textiles in that they are figured with images representing a variety of symbolic meanings that reflect religious beliefs and civil behavior held by their respective cultures.

As early as the Han Dynasty (202 B.C. to 220 A.D.), Chinese silk production had come under the control of state bureaucracy to assure that official dress codes would be strictly adhered to. The style, color and images on the official's robe indicated his status and authority in the political hierarchy. By the Song Dynasty (960-1279 A.D.) a new technique of weaving called k'o-ssu was introduced to China by the Uighur Turks. K'o-ssu is comparable to, but much finer than, Western tapestry weave, and only the very wealthy nobility could afford k'o-ssu. It was highly desired because of the beautiful and distinct images that could be woven with this technique.

The Japanese learned from the Chinese about sericulture and silk weaving. Although the Japanese copied the Chinese silk designs and style of dress over long periods of time, ultimately the Japanese adapted these designs and styles of dress to fit their particular taste.

The Lotus Collection exhibits and sells fine textiles not only from China and Japan but from European cultures as well. Located in the historic Jackson Square district of San Francisco, the Lotus Collection has in its extensive inventory textile art dating from the 2nd century through the late 19th century. Examples range in size from a few square inches to yardage, full costumes, and complete room-size tapestries.

INTRODUCTION TO SELECTIONS FROM *THE STORY OF THE STONE*

To bring to life the social world within which the exquisite textiles featured in this day book were created and used, the reproductions are interspersed with selections from a fascinating 18th century Chinese novel.

The Story of the Stone, also known as *Dream of the Red Chamber,* is a vivid and important document of social manners in old Chinese cultures.

The first of its five volumes, titled "The Golden Days," describes the glory and decline of the Jia Family. The life stories of its main characters, Bao-yu and Dai-yu, are presented in humorous ways and at the same time in realistic detail, reflecting the ritualized hurly-burly of Chinese high society of those days.

Its style is delicate like poetry, reflecting not only the stylishness of the lifestyle but also a deeper, metaphysical dimension, the constant awareness of the embeddedness of everyday life in another plane of existence, according to the Buddhist belief in a supernatural scheme of all things.

JANUARY

NEW YEAR'S DAY

1

2

3

4

5

6

7

A *Chinese 14th c. Yuan dynasty (1260–1368) embroidered red diaper patterned panel.* Embroidered images of two lotus blooms are executed in couched metallic gold wrapped threads, outlined in polychrome silk embroidery, with a lotus pad below and between each done in silk floss threads. Two mandarin ducks on the left and right swim on top of embroidered stylized waves which are highlighted with couched metallic gold threads. *(Courtesy of Neutrogena Corporation)*

Her chignon was enclosed in a circlet of gold filigree and clustered pearls. It was fastened with a pin embellished with flying phoenixes, from whose beaks pearls were suspended on tiny chains.

Her necklet was of red gold in the form of a coiling dragon.

Her dress had a fitted bodice and was made of dark red silk damask with a pattern of flowers and butterflies in raised gold thread.

Her jacket was lined with ermine. It was of slate-blue stuff with woven insets in colored silks.

Her under-skirt was of a turquoise-colored imported silk crêpe embroidered with flowers.

8

9

10

11

12

13

14

15

16

17

18

19

20

21

Detail of a 19th c. Chinese velvet uncut lady's jacket. The blue silk satin ground is figured with cut silk pile images of flowering tree peonies, peaches, and roundels encompassing a *Shou* (good fortune) character.

Here there was a large kang *underneath the window, covered with a scarlet Kashmir rug. In the middle of the* kang *was a dark-red bolster with a pattern of medallions in the form of tiny dragons, and a long russet-green seating strip in the same pattern.*

In the part of the room below the kang *there was a row of four big chairs against the east wall. All had footstools in front of them and chair-backs and seat covers in old rose brocade sprigged with flowers.*

22

23

24

25

26

27

28

29

30

31

1

2

3

4

Detail of a Chinese late Ming (1368–1644) column hanging. In this detail, the quality of the embroidery work can be seen, particularly the 'needle' shading executed on the scales of the dragon's body, each of which is then outlined in couched metallic gold wrapped threads. The extremely high quality of the workmanship employed in the creation of this piece suggests that it was created for Imperial use.

The young gentleman who entered in answer to her unspoken question had a small jewel encrusted gold coronet on the top of his head and a golden headband low down over his brow in the form of two dragons playing with a large pearl. He was wearing a narrow sleeved, full skirted robe of dark red material with a pattern of flowers and butterflies in two shades of gold. It was confined at the waist with a court girdle of colored silks braided at regular intervals into elaborate clusters of knotwork and terminating in long tassels.

Over the upper part of his robe he wore a jacket of slate blue Japanese silk damask with a raised pattern of eight large medallions on the front and with tasseled borders.

On his feet he had half length dress boots of black satin with thick white soles.

5

6

7

8

9

10

11

12

13

14

15

16

17

18

FEBRUARY

A Chinese 17th c. K'ang-hsi (1662–1722) k'o-ssu (silk tapestry weave) fragment. Woven with metallic gold wrapped threads and polychrome silk threads, the ground is figured with a state lantern motif containing a pair of mandarin ducks floating amidst scattered lotus leaves beneath a flowering prunus branch.

Quite soon he was back once more, this time dressed in a completely different outfit.

The crown and circlet were gone. She could now see that his side hair was dressed in a number of small braids plaited with red silk, which were drawn round to join the long hair at the back in a single large queue of glistening jet black, fastened at intervals from the nape downwards with four enormous pearls and ending in a jeweled gold clasp.

19

20

21

22

23

24

25

FEBRUARY / MARCH

26

27

28

LEAP DAY
29

1

2

3

A mid 18th c. Chinese tribute silk panel, with a detail of the reverse side showing the discontinuous weft threads of the brocading. The green silk ground is brocaded in polychrome silks and metallic gold wrapped threads with a repeated image of a celestial horseman carrying a piece of red coral, a symbol of longevity, interspersed with sacred objects.

He had changed his robe and jacket for a rather more worn looking rose colored gown, sprigged with flowers. He wore the gold torque and his jade as before, and she observed that the collection of objects round his neck had been further augmented by a padlock shaped amulet and a lucky charm. A pair of ivy colored embroidered silk trousers were partially visible beneath his gown, thrust into black and white socks trimmed with brocade. In place of formal boots he was wearing thick soled crimson slippers.

4

5

6

7

8

9

10

MARCH

11

12

13

14

15

16

ST. PATRICK'S DAY

17

Late 18th c. Chinese child's embroidered jacket on silk gauze figured with archaic dragons. The large central roundel is embroidered with auspicious flowering tree peonies, a lily, sacred rocks, bats, and the fungus of immortality in gold wrapped threads with highlights of silk floss threads on flowers and bats.

A dark-red patterned curtain hung from brass hooks over the doorway. Inside, under the window in the south wall, there was a kang *covered with a dark-red carpet. At the east end of the* kang, *up against the wooden partition wall, were a backrest and bolster, both covered in gold brocade, and a large flat cushion for sitting on, also glittering with gold thread.*

18

19

20

21

22

23

24

25

26

27

28

29

30

31

A Chinese 17th c. K'ang-hsi (1662–1722) k'o-ssu (silk tapestry weave) fragment. The coral red ground is figured with stylized clouds, a sacred rock, and a flying egret amongst lotus blossoms above a traditional wave pattern. *(Courtesy of Louise Woodhead, Antique Textiles and Oriental Carpets).*

Wang Xi-feng had on a little cap of red sable, which she wore about the house for warmth, fastened on with a pearl studded bandeau. She was dressed in a sprigged peach pink gown, with an ermine lined skirt of dark red foreign crepe underneath it, and a cloak of slate blue silk with woven colored insets and lining of grey squirrel around her shoulders.

1

2

3

4

5

6

7

APRIL

8

9

10

11

12

13

14

19th c. Japanese blue silk brocade kesa (Buddhist priest's robe) patterned with stylized white mums and gold swastikas (denoting good fortune). The low number of 'columns' (vertical divisions denoting rank) with which this piece is constructed indicates that the wearer was probably a novice monk.

She was wearing a honey colored padded gown, a mulberry colored sleeveless jacket with a pattern in gold and silver thread, and a greenish yellow padded skirt. All her clothing had the same sensible, rather well worn look about it.

15

16

17

18

19

20

21

22

23

24

25

26

27

28

Detail of previous picture. Detail of the embroidered brocade pattern with red swastikas (a Chinese emblem for good fortune) and a beautifully embroidered manchurian crane with the top of his head rendered in red silk thread executed in the Peking knot stitch, standing next to a rock with a flower growing out of it.

Looking up, Boa-yu saw that Shui Rong's princely headgear was embellished by way of mourning with white bands, a white hatpin, and filigree silver 'wings.' As a further token of mourning his robe, though heavily bordered with a 'tooth and wave' design of rainbow-colored stripes and gold-emblazoned with the royal five-clawed dragon, was of a white material. It was confined at the waist by a red leather belt, studded with green jade.

29

30

1

2

3

4

5

6

7

8

9

10

11

12

Late Ming (1368–1644) embroidery fragment. Executed in polychrome silk threads on a woven metallic gold cloth ground, this embroidery depicts the '100 Children' theme, all portrayed as boys. These boys carry auspicious Taoist symbols of longevity in demonstration of their filial piety to the seated scholar. Scattered throughout the composition are important Buddhist emblems. Because fertility is also associated with this theme, often this type of hanging was created for wedding celebrations.

Now a great dressing up began as each lady robed herself in the costume appropriate to her rank. Then off they went to the Palace in four sedans one behind the other: Grandmother Jia's at the head, then Lady Xing's, then Lady Wang's, and then You-shi's. Jia She and Cousin Zhen also changed into court dress, and taking Jia Rong and Jia Qiang with them, accompanied Grandmother Jia to the Palace as her male escort.

13

14

15

16

17

18

19

MAY

MAY

20

21

22

23

24

25

26

A late 18th or early 19th c. Japanese silk tapestry woven (tsusure-ori) wall hanging. The tan colored ground is figured with images of two *ikibana* baskets filled with flowers. The large basket is filled with peonies, chrysanthemums, lotus, iris, and a blossoming wisteria branch. The smaller basket is filled with paperwhites and mums.

"Our present Emperor, who has always had a great sympathy for the common man, believes that the filial affection of a child for its parents is the most important thing in the world, and that family feeling is the same everywhere, irrespective of social rank. He has found that in his own case, even after seeing the Ex-Emperor and Ex-Empress morning, noon and night every day of his life, he is still unable to express more than a fraction of the devotion he feels for them: and this has led him to think of all those concubines and maids of honor and other court ladies, taken from their homes and shut up in the Palace for years and years on end, and to realize how much they must miss their *parents. And from there he got to thinking of the parents themselves, how they must long for the daughters they can never see again. And then he thought what a crime against Nature it would be if any of those parents were to become ill as a result of not seeing their daughters. So in the end, by an act of supreme generosity, the Emperor issued a special decree in which he said that, apart from allowing the families of court ladies to visit their daughters in the Palace on the twelfth day of each month, he would allow any family which had a separate house or annex capable of being maintained in the degree of security specified for a Temporary Imperial Residence to make written application for permission to receive a Visitation in their own home, where the pleasures of reunion could be enjoyed in an atmosphere of intimacy and affection."*

27

28

29

30

31

1

2

JUNE

3

4

5

6

7

8

9

Detail of a Chinese 17th c. k'ang-hsi (1662–1722) embroidered cream colored silk panel. The panel from which this detail of a mythical bird amidst flower blossoms is taken is from a set of a lady's bed hangings embroidered in polychrome threads and highlighted with gold wrapped threads and images of flowers, birds, and butterflies.

In response to his inquiry Jia Lian extracted a wallet from the leg of his boot, and glancing at a folded schedule inside it, summarized its contents as follows:

"Curtains, large and small, in various silks and satins—flowered, dragon-spot, sprigged, tapestry, panelled, ink-splashed: one hundred and twenty. —Eighty of those were delivered yesterday. That leaves forty yet to come. —Blinds: two hundred. —Yes. They all arrived yesterday. But then there are the special ones. —Blinds, scarlet felt: two hundred. Speckled bamboo: one hundred. Red lacquered bamboo with gold fleck: one hundred. Black lacquered bamboo: one hundred. Colored net: two hundred. —We now have half of each of those four kinds. The other half is promised by the end of autumn. —Chair-covers, table drapes, valences, tablecloths: one thousand two hundred of each. —Those we already have."

10

11

12

13

14

15

16

17

18

19

20

21

22

23

A Chinese late 17th c. K'ang-hsi (1662–1722) fragment of an Imperial dragon robe. The metallic gold wrapped thread couched ground is embroidered with a dragon executed in gold wrapped threads, peacock feather wrapped threads, and polychrome silk threads. This dynamic and energetic dragon holds the shining pearl of happiness.

The trompe-l'oeil *effect of these ingenious partitions had been further enhanced by inserting false windows and doors in them, the former covered in various pastel shades of gauze, the latter hung with richly patterned damask portieres. The main walls were pierced with window-like perforations in the shape of zithers, swords, vases and other objects of virtue.*

24

25

26

27

28

29

30

1

2

3

INDEPENDENCE DAY

4

5

6

7

Late 18th or early 19th c. Japanese brocade. This brocade has stylized red lotus blooms, pods, and vines on a blue silk ground.

JULY

By five o'clock next morning, when it was still dark night outside, all those in the family from Grandmother Jia downwards who held any sort of rank or title were already dressed in full court rig. In the Separate Residence

painted phoenix and coiling dragon
flapped and fluttered on drapes and curtains,
gold and silver-work gleamed and glinted,
jewels and gems made a fiery sparkle,
subtle incenses smoldered in brazen censers,
'everlastings' blossomed in china vases;

—and all was so silent, that throughout the whole of that great garden not a sound of a cough was to be heard.

JULY

8

9

10

11

12

13

14

JULY

15

16

17

18

19

20

21

Transitional Ming (1368–1644) to Ch'ing (1644–1912) late 17th c. Chinese rank badge for a seventh degree official.
This badge features a Mandarin duck on a gold couched ground highlighted with couched peacock feather filaments and polychrome silk embroidery.

Presently a faint sound of music was heard and the Imperial Concubine's procession at last came into sight.

First came several pairs of eunuchs carrying embroidered banners.

Then several more pairs with ceremonial pheasant-feather fans.

Then eunuchs swinging gold-inlaid censers in which special 'palace incense' was burning.

Next came a great gold-colored 'seven phoenix' umbrella of state, hanging from its curved-topped shaft like a great drooping bell flower. In its shadow was borne the Imperial Concubine's travelling wardrobe: her head-dress, robe, sash and shoes.

Eunuch gentlemen-in-waiting followed carrying her rosary, her embroidered handkerchief, her spittoon, her fly-wisk, and various other items.

Last of all, when this army of attendants had gone by, a great gold-topped palanquin with phoenixes embroidered on its yellow curtains slowly advanced on the shoulders of eight eunuch bearers.

JULY

22

23

24

25

26

27

28

29

30

31

1

2

3

4

Detail of a Chinese 17th c. or late Ming (1368–1644) fragment from a wall hanging. This central medallion from a dark blue silk velvet wall hanging depicts a mythical beast called a *kylin* (unicorn), within an inner cloud form medallion of couched gold wrapped threads, surrounded by a field of auspicious lotus and mum blooms, bordered by a scrolling vine and leaf motif.

Because of the season, the willows and apricot trees above them were bare and leafless; but in place of leaves they were festooned with hundreds of tiny lanterns, and flowers of gauze, rice paper and bast had been fastened to the tips of their branches. Other lanterns of shells and feathers, in the form of lotuses, water lilies, ducks and egrets floated on the surface of the water below.

5

6

7

8

9

10

11

AUGUST

12

13

14

15

16

17

18

Chinese 17th c. K-ang-hsi (1644–1722) Chinese lady's informal robe fragment. This robe of celadon silk is beautifully embroidered with a brocade pattern surrounding a roundel which encompasses images of two phoenixes and blossoming peonies.

On the day following the Imperial Concubine's return to the palace, she called upon the Emperor to offer thanks and gave him a full account of her Visitation. The Emperor was visibly pleased by her report and commanded that bounties of gold, silver and silks should be issued to Jia Zheng and the fathers of visiting ladies by the Inner Treasury.

19

20

21

22

23

24

25

26

27

28

29

30

31

1

A late 16th c. Chinese embroidered wall hanging in a woven metallic gold ground of a birthday celebration in an elaborate garden setting. At the center of the composition is the person being honored, seated upon a throne-like chair, within a garden pavilion. Servants and guests bear gifts for the occasion, while musicians and dancers are in the foreground entertaining. Court ladies wander through the environs of the very large garden.

Against the wall beneath was a long, high table. On it, towards the left, stood a large Northern Song porcelain dish heaped with those ornamental citrus fruits they call 'Buddha's hands,' whose brilliant yellow contrasted agreeably with the greenish blue of the glaze . . . On the opposite side of the room was a large four-poster bed whose silk gauze hangings had a pattern of bright green plants and insects in reversible embroidery.

2

3

4

5

6

7

8

9

10

11

12

13

14

15

Detail of an early 18th c. Japanese brocade panel. The navy blue ground silk is brocaded with images of large flower heads and vines in a repeated pattern highlighted with gold threads.

At that moment Bao-qin arrived. She was wearing a magnificent rain cape that glittered as she moved with gold and greenish lights. Bao-chai asked her where she had got it from.

"Lady Jia gave me it," said Bao-qin. "She looked it out for me because it was beginning to sleet."

Caltrop examined it curiously.

"No wonder it looks so beautiful: this is woven out of peacock's down."

"That's not peacock's down," said Xiuang-yun. "It's made from mallard's head feathers." She smiled at Bao-qin teasingly: "One can see how fond of you she must be. She's fond of Bao-yu, but she's never let him wear this."

16

17

18

19

20

21

22

23

24

25

26

27

28

29

Detail of the top half of a Chinese 17th c. K'ang-hsi (1662–1722) metallic gold ground k'o-ssu (silk tapestry weave) chair panel. This chair panel is woven in polychrome silks figured with images of pheasants perched upon sacred rock formations, flowering tree peony branches, and at the top a *Shou* symbol (a longevity symbol) executed in archaic script amidst stylized clouds.

"Is it snowing?" she asked him.

"Not yet," said Bao-yu, "but it looks as if it will."

Grandmother Jia turned to Faithful:

"Give him that peacock feather cloak we were looking at yesterday."

Faithful murmured a reply and went out of the room, returning presently with a magnificent snow cape that gleamed and glittered with gold and green and bronzy bluish lights. It was like Bao-qin's mallard cape and yet somehow different.

"This is what they call 'peacock gold,' " said Grandmother Jia. "It is made by the Russians. They twist the barbs of peacock feathers into a kind of yarn and weave it from that. The other day I gave your cousin Bao-qin a cape like this made out of mallard feathers."

30

1

2

3

4

5

6

7

8

9

10

11

12

13

Detail of the bottom half of the chair panel shown in the previous picture. Shown on a metallic gold ground are images of auspicious Buddhist symbols, clouds, and a side-facing dragon above a rock and wave motif.

And so, before taking their places on the kang, *they went off to remove their hair ornaments and make themselves more comfortable. They returned wearing only tunics and trousers, with their unadorned hair loosely knotted or coiled. Bao-yu himself was wearing a dark red cotton tunic tied at the waist with a sash, and trousers in a black and green patterned lined silk gauze, unconfined at the ankles.*

14

15

16

17

18

19

20

21

22

23

24

25

26

27

Detail of an 18th c. Chinese temple banner. This banner is embroidered in polychrome silk threads on a blue silk satin ground with an image of a phoenix or *feng-huang,* the second of the four miraculous creatures. The phoenix traditionally represents the female side of nature in Chinese art and is often used on wedding robes and hangings created for use in women's quarters.

The girls found him already ensconced on the kang. *He was leaning back, one elbow resting on a newly made turquoise colored pouffe stuffed with rose and peony petals, playing guess fingers with Parfumée. Parfumée was shouting excitedly as she played. She was wearing a very short, close fitting tunic with a harlequin pattern of turquoise, deep purple and reddish brown lozenges, a willow green sash, and flower sprigged pale red trousers which, like Bao-yu's, were unconfined at the ankles. Her hair was done up in a number of little plaits which had been drawn back to join one great plait, as thick as a goose's egg at the nape of her neck.*

OCTOBER/NOVEMBER

28

29

30

HALLOWEEN
31

1

2

3

4

5

6

7

8

9

10

Detail of 19th c. Chinese red silk ground mirror cover. This work is embroidered with the image of a male figure standing on a sea dragon's head delivering a flower to the shore.

Out of deliberate disregard for appearances she had taken off her hair ornaments and outer clothes, and from time to time as she spoke, the animated gestures with which her words were accompanied caused the imperfectly fastened shift she was wearing to gape open, revealing glimpses of leek green breast binder and snow white flesh beneath; the red shoes that peeped out below her green drawers were all the time tap tapping or coming together in a manner that was anything but ladylike, and her earrings bobbed to and fro like little swings.

VETERANS' DAY

11

12

13

14

15

16

17

NOVEMBER

18

19

20

21

22

23

24

Late 18th c. Japanese blue ground silk tapestry (tsuzure-ori). The central roundel encompasses mums, peonies, prunus branches, and a mythical bird of paradise. The four corners are figured with pairs of fowl, doves, and sparrows and manchurian cranes, each accompanied by a spray of flowers or branches. Metallic gold wrapped threads have been woven in throughout this piece to add highlights.

To her brow's dusky crown and lips incarnadine, the lamplight lent an added softness and brightness; and the wine that she drunk gave her eyes, which were at all times sparkling and vivacious, an even more irresistible allure. The two men sat spellbound, and at the same time repelled.

25

26

27

28

29

30

1

DECEMBER

2

3

4

5

6

7

8

Chinese late 17th c. K-ang-hsi (1662–1722) k'o-ssu (silk tapestry weave) ceremonial cape. The yellow ground of this extremely rare item is figured with scrolling foliate vines, embellished with lotus leaves and lotus flower heads, all within a red border containing elements of the celestial landscape and emblems of good fortune, highlighted with metallic gold threads which have been woven into the composition.

He observed that she was wearing a little pale blue fur-lined dress embroidered with flowers, and an ermine-lined sleeveless jacket, while her hair was coiled up in her everyday style and had no flowers in it but only a flat hairpin of purest gold. Her padded underskirt was pink, and embroidered with flowers. How graceful she seemed, as a jade tree leaning in the wind; how gentle, as a fragrant lotus whose petals are moist with dew.

9

10

11

12

13

14

15

DECEMBER

16

17

18

19

20

21

22

Detail of an early Ming Dynasty (1368–1644) 15th c. Chinese k'o-ssu (silk tapestry weave) hanging. This piece is figured with stylized horizontal bands of clouds across which pairs of auspicious birds fly.

"Who's there?" Xue Ke called out.

No answer. He got out of bed to open the door and saw Moonbeam, her hair combed simply back, wearing a tight little sleeveless jacket, its gold-striped neck-line cut in the shape of a guitar, its buttons seductively undone. She had a new-looking viridian scarf around her neck and instead of a skirt was sporting pomegranate-red lined pantaloons of a flowery design, and a pair of smart red embroidered slippers. She had come before doing her morning toilet, to remove the sweetmeats before the rest of the household arose. Though rather taken aback to see her advancing into his room en néglIgé, *he asked politely: "Why are you up so early this morning?"*

Moonbeam blushed in reply, piled the sweetmeats onto one plate and went out holding it in both hands. Xue Ke interpreted this as pique at her failure to seduce him the previous evening.

DECEMBER

23

24

CHRISTMAS
25

26

27

28

29

30

31

Chinese 17th c. gold ground k'o-ssu (silk tapestry weave) fragment. The yellow citron in a bowl on a stand is beautifully detailed with blue and white mixed silk threads and surmounted by a vase patterned with geometric designs and filled with a blossoming peony. This finger-like citron is also called 'Buddha's hand'. It has a very strong fragrance and is often put in porcelain bowls and placed on the altar of the household gods at the New Year Festival.

"Where is the other casket?" asked Feng turning to his page, who promptly came forward with a rosewood casket held aloft in both hands. The three men gathered round as it was opened. On its lining of tiger-brocade lay a length of blue gauze-like material, many times folded.

"And what is this?" asked Zhan Guang.

"This," replied Feng, "is called Byssus Net."

He took it from the casket and laid it on the table. Folded as it was, it occupied a space no more than five inches long and less than half an inch thick. Feng began to unfold it. When he had done so a dozen times, it extended over the edge of the table.

"There are still two folds to come," he explained. "To unfold it to its full extent we would need to hang it in a room with a high ceiling. This fabric is woven from the Byssus, the so-called Mermaid's Tears. In extreme heat it would make a perfect fly and mosquito-net for use in a large reception hall. As you can see, it is extremely light and transparent."